Are You A Shwappy Fwapperoo?

A Children's Anytime Story by
Greg Horn and Devon Horn

Illustrated by
Lisa Eiler

Copyright 2009
All Rights Reserved

AF883615

Created Just For You
By

Greg Ho

Devon

lisa eiler

Are you a Shwappy Fwapperoo?

Or does a Skeeziks look like you?

COOKIES
DOG

Or maybe you're
a Whizzlezitz.

With dwoopy
clothes that
barely fit.

Are you a Skatarillakat?

Or do you just sometimes act like that?

Are you a
Stinky Winkerbean?

Does asking
you make me
seem mean?

SEEDS
SOIL

Are you a funny Shummybunny?

Who fwaps outside when it is sunny?

MEMOIRS
BUNNY DIGEST

Or just a
Shwippy Skizzelzoo?

If so, can I have
fun with you?

Are you a Piffy Pagabug?

Just waiting for a great big hug?

Are you a
Shwappy
Fwapperoo?

I think you are.

And I'm
one too!

Are You a Shwappy Fwapperoo?

About the Authors

Greg Horn (also known as Dadn at home) lives with his family in Lighthouse Point, Florida. This book was a team project with his twin daughter Devon, who began making up funny words when she was two. Dadn put them into story form, and this children's anytime book was born.

About the Illustrator

Lisa Eiler is a fine artist and illustrator living in southern New Jersey. She specializes in children's illustration and wildlife art. She is an avid bird watcher and finds the pine barrens of New Jersey to be a paradise that supports her passion for art and nature.

Made in United States
Cleveland, OH
05 November 2026

47649174R00017